作者简介 | AUTHOR PROFILE

叶 斌 Ye Bin

高级建筑师
国广一叶装饰机构首席设计师
福建农林大学兼职教授
南京工业大学建筑系建筑学学士
北京大学 EMBA
中国建筑学会室内设计分会理事
中国建筑装饰协会理事

Senior Architect
Chief Architect of Guoguangyiye Decoration Group
Adjunct Professor of Fujian Agriculture and Forestry University
B. Arch from Nanjing Industry University
EMBA from Beijing University
Councilor Member of China Institute of Interior Design
Councilor Member of China Building Decoration Association

荣誉

荣获"中国室内设计杰出成就奖"
当选 2009 年"金羊奖"中国十大室内设计师
当选中国建筑装饰行业新中国成立 60 年百名功勋人物
当选 1989~2009 年中国杰出室内设计师
当选 1997~2007 年中国家装十年最具影响力精英领袖
当选 1989~2004 年全国百位优秀室内建筑师
当选 2004 年度中国杰出中青年设计师
当选 2004 年度中国室内设计师十大封面人物
当选 2002 福建省室内设计十大影响人物(第一席位)

著作

1. 《室内设计图典》(1、2、3)
2. 《装饰设计空间艺术·家居装饰》(1、2、3)
3. 《装饰设计空间艺术·公共建筑装饰》
4. 《建筑外观 Appearance 细部图典》
5. 《国广一叶室内设计模型库·家居装饰》(1、2、3)
6. 《国广一叶室内设计模型库·公建装饰》
7. 《国广一叶室内设计》
8. 《国广一叶室内设计模型库构成元素》(1、2)
9. 《室内设计立面构图艺术》系列
10. 《国广一叶室内设计模型库》系列
11. 《家居装饰·平面设计概念集成》
12. 《概念家居》
13. 《概念空间》
14. 《国广一叶家居装饰》系列
15. 《室内设计图像模型》系列
16. 《2009 室内设计模型》系列(5 册)
17. 《2010 家居空间模型》系列(3 册)
18. 《2010 公共空间模型》系列(2 册)
19. 《2011 家居空间模型》系列(3 册)
20. 《2011 公共空间模型》
21. 《2012 室内设计模型集成》系列(5 册)
22. 《名家家装 + 材料标注》系列(5 册)
23. 《2013 家居空间模型集成》系列(3 册)
24. 《2013 公共空间模型集成》系列(2 册)
25. 《2014 家居空间模型集成》系列(4 册)
26. 《2014 公共空间模型集成》
27. 《2015 室内设计模型集成》系列(共 5 册)
28. 《名师家装新图典》系列(3 册)

获奖设计作品

叶禅赋	2015 年第十八届中国室内设计大奖赛银奖
溪山温泉度假酒店(实例)	2014 年第十届中国国际室内设计双年展金奖
正兴养老社区体验中心	2014 年第十届中国国际室内设计双年展银奖
中联大厦办公楼	2014 年第十届中国国际室内设计双年展银奖
尊贵彰显富丽	2014 年第十届中国国际室内设计双年展铜奖
永福设计研发中心	2014 年度全国建筑工程装饰奖(公建筑装饰设计类)
宇洋中央金座	2013 年第十六届中国室内设计大奖赛铜奖
书·韵	2013 年国际空间设计大奖"Idea-Tops 艾特奖"提名奖
聚春园驿馆	2013 年国际空间设计大奖"Idea-Tops 艾特奖"入围奖
童话地中海	2013 年国际空间设计大奖"Idea-Tops 艾特奖"入围奖
宇洋中央金座	2013 年国际空间设计大奖"Idea-Tops 艾特奖"入围奖
宁德上东曼哈顿售楼部	2013 年第四届中国国际空间环境艺术大赛(筑巢奖)优秀奖
福建洲际国际酒店	2012 年第二届亚太酒店设计大赛金奖
前线共和广告	2012 年第十五届中国室内设计大奖赛金奖
瑞莱春堂福州三坊七巷店	2012 年"照明周刊杯"中国照明应用设计大赛一等奖
前线共和广告	2012 年第九届中国国际室内设计双年展金奖
福州情·聚春园	2011 年第九届中国国际室内设计双年展银奖
宁化世界客家文化交流中心	2011 年第九届中国国际室内设计双年展银奖
一信(福建)投资	2011 年第十四届中国室内设计大奖赛金奖
福建科大永合医疗机构	2011 年中国最成功设计大奖最成功设计奖
连江贵安海峡文化村酒店	2011 年中国(上海)设计节"金外滩"最佳概念设计奖
素丽娅泰水疗会所	2010 年第八届中国室内设计双年展金奖
摩卡小镇销售楼中心	2010 年第八届中国室内设计双年展银奖
大洋鹭洲	2010 年第八届中国室内设计双年展铜奖
素丽娅泰水疗会所	2010 年亚太室内设计双年展大奖赛商业空间设计银奖
中联江滨御景会所	2010 年亚太室内设计双年展大奖赛商业空间设计优秀奖
繁都魅影	2010 年亚太室内设计双年展大奖赛住宅空间设计银奖
繁都魅影	2010 年亚洲室内设计大奖赛铜奖
中央美苑	2010 年海峡两岸室内设计大赛金奖
繁都魅影	2010 年海峡两岸室内设计大赛金奖
光·盒中盒	2010 年海峡两岸室内设计大赛金奖
中联江滨御景会所	2010 年海峡两岸室内设计大赛银奖
皇帝洞书院	2009 年"尚高杯"中国室内设计大奖赛二等奖(全国商业类第三名)
北湖皇帝洞景区会所	2008 年第七届中国室内设计双年展金奖
国广一叶点房财富中心	2007 年福建省室内设计大奖赛一等奖(公建工程类第一名)
国广一叶大家会馆	2006 年福建省室内设计大奖赛一等奖(公建工程类第一名)
点房财富中心	2007 年"华耐杯"中国室内设计大奖赛二等奖(全国商业类第二名)
大家会馆	2006 年第六届中国室内设计双年展金奖
书香大销售中心	2006 年第六届中国室内设计双年展银奖
金钻世家某单元房	2006 年第六届中国室内设计双年展银奖
福州金龙门餐厅	2006 年第六届中国室内设计双年展银奖
滨江丽景·美丽园	2006 年第六届中国室内设计双年展银奖
福建电力调度通信中心大楼	2006 年第六届中国室内设计双年展铜奖
金源国际酒店桑拿中心	2006 年第六届中国室内设计双年展优秀奖
内蒙古呼和浩特市中级人民法院	2004 年第五届中国室内设计双年展铜奖
厦门奥林匹亚中心	2004 年第五届中国室内设计双年展铜奖

叶 猛 Ye Meng

国广一叶装饰机构副总设计师
国家一级注册建筑师
国家一级注册建造师
中国建筑学会室内设计分会会员
福建工程学院建筑与规划系讲师
福州大学建筑系学士
中南大学土建学院建筑学硕士

Deputy Chief Architect of Guoguangyiye Decoration Group
First-Class Registered Architect (PRC)
Registered Constructor (PRC)
Member of Institute of Interior Design of Architectural Society of China
Lecturer of Architecture and Planning Dept., Fujian University of Technology
B. Arch from Fuzhou University
M. Arch from Central South University

获奖设计作品

融信大卫城	2014 年第十届中国国际室内设计双年展优秀奖
三盛国际公园	2014 年第五届中国国际空间环境艺术设计大赛(筑巢奖)提名奖
名城港湾	2014 年第五届中国国际空间环境艺术设计大赛(筑巢奖)优秀创意奖
融侨外滩	2014 年第五届中国国际空间环境艺术设计大赛(筑巢奖)优秀创意奖
鳌峰洲小区—19A	2013 年第四届中国国际空间环境艺术设计大赛(筑巢奖)优秀奖
阳光理想城	2011 年第九届中国国际室内设计双年展金奖
大洋鹭洲	2010 年第八届中国室内设计双年展铜奖
繁都魅影	2010 年亚洲室内设计大奖赛铜奖
福建工程学院建筑系新馆	2009 年中国室内空间环境艺术设计大赛一等奖
福建工程学院建筑系新馆	2009 年福建室内与环境设计大奖赛公建工程类最高奖
文化主题酒店	2008 年福建省第六届室内与环境设计大赛一等奖
旗山文城	2008 年福建省第六届室内与环境设计大赛一等奖
另类博弈	2008 年第六届现代装饰年度办公空间大奖入围奖
点房财富中心	2007 年"华耐杯"中国室内设计大奖二等奖
翻阅古朴	2007 年福建省第五届室内设计与环境大赛一等奖
大家会馆	2006 年第六届中国室内设计双年展金奖
福建电力调度通信中心大楼	2006 年第六届中国室内设计双年展铜奖

另出版《建筑外观细部图典》、《室内设计图像模型》等著作数十种

前言 / PREFACE

国广一叶装饰机构，作为"全国最具影响力室内设计机构"（中国建筑学会室内设计分会颁发）、"2014年度中国建筑装饰设计机构50强企业"（中国建筑装饰协会颁发）、"2013住宅装饰装修行业最佳设计机构"（中国建筑装饰协会颁发）、"2013年度全国住宅装饰装修行业百强企业"（中国建筑装饰协会颁发）、"2012～2013年度全国室内装饰优秀设计机构"（中国室内装饰协会颁发）、"2012年中国十大品牌酒店设计机构"（中外酒店论证颁发）、"2011～2012年度全国室内装饰优秀设计机构"（中国室内装饰协会颁发）、"1989～2009年全国十大室内设计企业"（中国建筑协会室内设计分会颁发）、"1988～2008年中国室内设计十佳设计机构"（中国室内装饰协会颁发）、"1997～2007年中国十大家装企业"（中国建筑装饰协会颁发）、"福建省著名商标"、"省、市级重合同守信用企业"、"福建省建筑装饰装修行业龙头企业"（福建省人民政府闽政文〔2014〕26号颁发），"福建省建筑装饰行业协会会长单位"，荣获国际、国家及省市级设计大奖上千项。

国广一叶装饰机构首席设计师叶斌，曾荣获"中国室内设计杰出成就奖"、两次荣获"中国十大室内设计师"称号；叶猛被评为"1989～2009年中国优秀设计师"；另外，19位设计师被评为中国装饰设计行业优秀设计师，83位设计师分别被评为福建省优秀设计师、福州市优秀设计师，76位在职设计师分别荣获历届全国、福建省、福州市室内设计一等奖。

以上这些荣誉的获得，是和国广一叶装饰机构自身的水准密切相关的。国广一叶装饰机构拥有大批量高水准的室内设计师设计的专业效果图，这些效果图将设计师的设计意图淋漓尽致地表现出来。

自2004年国广一叶装饰机构在福建科学技术出版社出版了第一套模型系列图书，深得读者青睐后，每年出版一套，至今已陆续出版了12套模型系列图书，一直受到广大读者的支持与厚爱。现继续推出的《2016公共空间模型库》和《2016家居设计模型库（共4种）》系列，精选的是国广一叶装饰机构在2015至2016年间设计的最新作品，汇集了1700多个风格各异、手法时尚的室内设计效果图及其对应的3ds Max场景模型文件，可作为读者做室内设计时的有益参考。

本书配套光盘的内容包含效果图及其原始3ds Max模型和使用到的所有贴图文件。由于3ds Max软件不断升级，此次的模型我们采用3ds Max2014版本制作。模型按图片顺序编排，易于查阅调用。只有能对模型进一步调整才能体现其价值和生命力，因此提供的3ds Max模型是真正有价值、可随时提取调整用的部分。必须说明的是，书中收录的效果图均为原始模型经过lightscape渲染和photoshop后处理过的成图，是为读者了解后处理效果提供直观准确的参考，与3ds Max直接渲染的效果有一定的区别。

著 者
2016年2月

As a well-known decoration company, Guoguangyiye Decoration Group have acquired thousands of international, national and provincial design awards, such as "Top 50 architectural decoration company in China(2012, honored by China Building Decoration Association, CBDA)", "Outstanding Interior Design Companies in China(2011-2012 and 2012-2013, honored by China National Interior Decoration Association, CIDA)", "Top 10 Candlewood Design Companies in China(2012, honor by Chinese and Foreign Hotel Argument)," "The Best Interior Decoration Association of Chinese Home Decoration(2013, honored by CBDA), "Top 10 Interior Design Companies in China (1989~2009)", "Top 10 China Interior Design Institutions (1988~2008)", "2012 China top 10 Hotel Design Institutions", "China Top 10 Home Decoration Enterprises (1997~2007)" and "Well-Known Brand of Fujian".

In Guoguangyiye Decoration Group, a dozen of architects have be granted as "National 19 architect of China", and 83 architects have awarded as "Excellent Architect of Fujian province/Fuzhou", 76 architects have won top prize of national, Fujian provincial or Fuzhou. The chief architect Mr. Bin Ye has wined the award of "Distinguished Achievement Award of Chinese Interior Design", and awarded twice "China Top 10 Interior Design Architect". Mr. Meng Ye was awarded "Outstanding Architect of China (1989~2009)".

Naturally these achievements have been accomplished because of the high level interior designs of Guoguangyiye, but obviously cannot be attained without high level professional effect drawing that presents the design intent of architects incisively and vividly. Therefore as a product of the collective efforts of architect and graphic designer, it is closely related to the success of project design.

Since 2004, Guoguangyiye has published eleven series of books on design model database with Fujian Science and Technology Press and all of them have gained wide popularity by their richness and practicality. Therefore, this year we will continue to publish 2016 Home Spale Models Integration and 2016 Public Spale Models Integration. This new series consists of over 1700 chic 3ds Max scenario models of various style interior designs created by Guoguangyiye during 2015~2016. Being a model database, they could also be used as beneficial references for interior design.

The enclosed CD contains original 3ds Max models of decoration effect drawings and all the map files used in order to create them. Due to the continuous upgrading of 3ds Max software, version 2014 was adopted in the drawing of these models which are arranged in the order of the pictures to make them easily accessible. Since as only models that can be further adjusted are valuable, the 3ds Max moulds provided are all of true value and readily available. It should be noted that, all the effect drawings in the books are pictures rendered by lightscape and dealt with by Photoshop, to give an intuitive and precise reference for readers on the after effects which are different from those rendered directly by 3ds Max.

February 2016

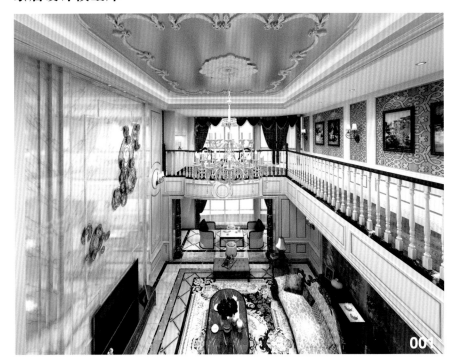

008-009
EUROPEAN STYLE
欧式风格

013

014

015

019

020

021

023

024

025

027

026

028

029

032

033

036

037

038

039

040

042

043

044

052

053

056

057

060

061

068

069

070

071

072

042-043
EUROPEAN STYLE
欧式风格

079

078

080

081

084

085

086

087

091

092

093

094

095

097

098

107

108

111

112

113

114

115

116

117

118

119

120

127

128

131

132

138

139

141

140

142

143

146

147

152

153

154

155

158

159

160

161

162

166

167

174

175

178

179

183

180

182

184

187

188

189

190

191

194

195

196

197

198

199

202

203

214

215

216

217

218

219

220

223

224

227

228

231

232

233

235

236

237

240

241

242

244

247

248

251

252

253

254

255

256

257

258

259

260

261

262

263

264

265

266

268

269

270

271

274

275

278

279

281

280

282

283

286

287

148-149
EUROPEAN STYLE
欧式风格

302

303

306

307

308

311

312

315

316

317

318

319

328

329

330

331

332

333

337

338

342

343

347

348

图书在版编目（CIP）数据

2016家居设计模型库.欧式风格/叶斌，叶猛著.—福州：福建科学技术出版社，2016.6
 ISBN 978-7-5335-4980-0

Ⅰ.①2… Ⅱ.①叶…②叶… Ⅲ.①住宅-室内装饰设计-图集 Ⅳ.① TU241-64

中国版本图书馆CIP数据核字（2016）第062675号

书　　名	2016家居设计模型库　欧式风格
著　　者	叶斌　叶猛
出版发行	海峡出版发行集团
	福建科学技术出版社
社　　址	福州市东水路76号（邮编350001）
网　　址	www.fjstp.com
经　　销	福建新华发行（集团）有限责任公司
印　　刷	恒美印务（广州）有限公司
开　　本	635毫米×965毫米　1/8
印　　张	22
图　　文	176码
版　　次	2016年6月第1版
印　　次	2016年6月第1次印刷
书　　号	ISBN 978-7-5335-4980-0
定　　价	268.00元

书中如有印装质量问题，可直接向本社调换